C000110573

ISBN 978-0-265-80753-8
PIBN 10892499

For support please visit www.forgottenbooks.com

Bibliographic Notes / Notes techniques et bibliographiques

to obtain the best original
features of this copy which
que, which may alter any of
oduction, or which may
sual method of filming are

L'Institut a microfilmé le meilleur exemplaire qu'il lui a
été possible de se procurer. Les détails de cet exemplaire qui sont peut-être uniques du point de vue bibli-
ographique, qui peuvent modifier une image reproduite,
ou qui peuvent exiger une modification dans la métho-
de normale de filmage sont indiqués ci-dessous.

ée

r laminated /
t/ou pelliculée

titre de couverture manque

s géographiques en couleur

r than blue or black) /
autre que bleue ou noire)

r illustrations /
tions en couleur

rial /
cuments

e

e shadows or distortion along
liure serrée peut causer de
orsion le long de la marge

ring restorations may appear
ver possible, these have been
l se peut que certaines pages
lors d'une restauration
exte, mais, lorsque cela était
ont pas été filmées.

☐ Coloured pages / Pages de couleur

☐ Pages damaged / Pages endommagées

☐ Pages restored and/or laminated /
Pages restaurées et/ou pelliculées

☑ Pages discoloured, stained or foxed /
Pages décolorées, tachetées ou piquées

☐ Pages detached / Pages détachées

☑ Showthrough / Transparence

☐ Quality of print varies /
Qualité inégale de l'impression

☐ Includes supplementary material /
Comprend du matériel supplémentaire

☐ Pages wholly or partially obscured by errata slips,
tissues, etc., have been refilmed to ensure the best
possible image / Les pages totalement ou
partiellement obscurcies par un feuillet d'errata, une
pelure, etc., ont été filmées à nouveau de façon à
obtenir la meilleure image possible.

☐ Opposing pages with varying colouration or
discolourations are filmed twice to ensure the best
possible image / Les pages s'opposant ayant des
colorations variables ou des décolorations sont
filmées deux fois afin d'obtenir la meilleure image
possible.

1	2	3

MICROCOPY RESOLUTION TEST CHART

(ANSI and ISO TEST CHART No. 2)

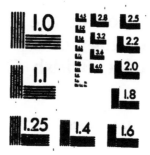

APPLIED IMAGE Inc.

1853 East Main Street
Rochester, New York   14609   USA
(716) 482 - 0300 - Phone
(716) 288 - 5989 - Fax

CANADA

DEPARTMENT OF MINES

HON. MARTIN BURRELL, MINISTER; R. G. McCONNELL, DEPUTY MINISTER.

GEOLOGICAL SURVEY

WILLIAM McINNES, DIRECTING GEOLOGIST.

MEMOIR 110

No. 89, GEOLOGICAL SERIES

Preliminary Report on the Economic Geology of Hazelton District, British Columbia

BY

J. J. O'Neill

OTTAWA
J. DE LABROQUERIE TACHÉ
PRINTER TO THE KING'S MOST EXCELLENT MAJESTY
1919

No. 1736

CANADA
DEPARTMENT OF MINES
Hon. Martin Burrell, Minister; R. G. McConnell, Deputy Minister.

Looking southeast towards Rocher D

CANADA
DEPARTMENT OF MINES
Hon. Martin Burrell, Minister; R. G. McConnell, Deputy Minister.

GEOLOGICAL SURVEY
William McInnes, Directing Geologist.

MEMOIR 110

No. 80, Geological Series

Preliminary Report on the Economic Geology of Hazelton District, British Columbia

BY

J. J. O'Neill

OTTAWA
J. de LABROQUERIE TACHÉ
PRINTER TO THE KING'S MOST EXCELLENT MAJESTY
1919

No. 1736

CANADA
DEPARTMENT OF MINES
GEOLOGICAL SURVEY

Preliminary Report on the Economic
Geology of
Hazelton District, British Columbia

CONTENTS.

Illustrations.

Preliminary Report on the Economic Geology of Hazelton District, British Columbia.

GENERAL STATEMENT AND ACKNOWLEDGMENTS.

The investigation of the economic geology of Hazelton district was undertaken in the summer of 1917 with a view to stimulating the mineral production from this easily accessible and promising district by determining the origin, distribution, and general occurrence of the ores, and their relations to one another, to the gangue minerals, and to the various country rocks which enclose the veins.

A topographical survey of the district was carried on at the same time by F. S. Falconer of the Topographical division, for a map on a scale of 1 mile to 1 inch with 100-foot contour intervals. The completed portion of this map is used as a base for the geological map.

The work was greatly facilitated by the interest and assistance of all the mining men in the district, and acknowledgment is made particularly to W. G. Norrie-Lowenthal of the Silver Standard, Messrs. Harris of the Harris mines, D. W. Williams of the Rocher De Boule, H. E. Clement of the Delta, and D. B. Morkill of the Hazelton View, for their courtesy and assistance during the examination of the properties under their management.

V. Dolmage, as field assistant, gave able and enthusiastic service which is being continued in the preparation of the final report, and H. J. James was a capable student assistant.

PREVIOUS WORK.

Geological reconnaissance work in the Hazelton district was done by W. W. Leach in 1909 and 1910 while preparing a topographic map of the region; his report is in the Summary Reports of the Geological Survey for those years.

In 1911 W. Fleet Robertson made a detailed report on the mining properties in the district, and the work has been elaborated and brought up to date from time to time since then by J. D. Galloway and the results published in the annual reports of the British Columbia Bureau of Mines from 1911 to 1917. In these reports the equipment and general development of the properties, analyses of the ores and their general occurrence are described.

In 1912 G. S. Malloch made a cursory examination of the ore deposits in the vicinity of Hazelton in connexion with his work in the Groundhog region to the north, and in the same year R. G. McConnell examined a geological section along the Grand Trunk Pacific railway through the district. Reports of these examinations are published in the Summary Report of the Geological Survey for 1912.

SUMMARY AND CONCLUSIONS.

There are several properties in the silver-lead-zinc group, and in the copper-gold group, which are not being worked, and which have showings indicative of a considerable amount of milling-ore and more or less high grade ore. These properties have been partly developed, but work on them has been shut down from lack of capital, the cost of mining and shipping the high grade material being prohibitive without returns from the milling-ore which must be opened up or extracted in mining the high grade ore.

For example, the claims on the north face of Ninemile mountain form a blanket which covers the whole upper slope and extends for 2 miles horizontally. These claims are divided into small groups under separate ownership, and at present none of them is being worked. The same veins extend across several claims or groups, and milling-ore is common to all the groups, but only a few have outcrops of high grade or shipping-ore sufficiently large to tempt prospectors to exploit the property for that alone. The conditions in the area are such that the whole western half at least could be worked—from one camp—as one mine with a common tram system. An aerial tram, less than 2 miles in length, would connect the workings with a mill-site on the Shegunia river where the power necessary for operating the mines could be developed.

On Rocher Déboulé mountain, especially in the Juniper Creek basin, the Rocher De Boule mine is the only property which is in a position to carry on mining to the best advantage. The various properties, though not favourably situated for complete co-operation, would be immensely benefited by co-operating in securing a central power-station, a custom mill located on Juniper creek, and a common tram system for carrying their concentrates and shipping ore to the railway. A good system of accounting would readily determine the proportion of cost of operation to be borne by each property.

In conclusion, it may be said that the district about Hazelton offers an excellent example of mining conditions over which a certain amount of government supervision would be of great advantage to the properties concerned, and to the country as a whole. If general co-operation were made compulsory in cases like this, it would tend to more efficient development of the district as a whole, and ensure a thorough test of all likely properties at a minimum of expense. It would also conserve a large amount of valuable mineral in lower grade properties, which under present conditions will be wasted, or rendered unavailable for extraction at a profit by the exhaustion of the more valuable properties.

All properties in the district are within reasonable distance of railway transportation; there is an abundance of water-power and of timber for mining; the veins are strong and contain deep vein minerals, and, therefore, may be expected to hold their values to considerable depth; the climate is good, and the labour market fair. With all these advantages it seems unwarrantable that properties should remain unproved.

GENERAL CHARACTER OF DISTRICT.

LOCATION AND AREA.

The district under investigation is known as the Hazelton sub-district of the Omineca mining division; it has an area of approximately 225 square miles, and is situated 130 miles northeast of Prince Rupert, or about 180 miles by rail. The confluence of the Skeena and Bulkley rivers at Hazelton lies in the northwest quadrant of the area. The district includes the mineral claims on Ninemile, Fourmile, Glen, and Rocher Déboulé mountains.

ACCESSIBILITY.

The Grand Trunk Pacific railway practically bisects the district in making a 90 degree bend around the Rocher Déboulé mountains, giving good shipping facilities to the different properties. The main trunk wagon roads of the Skeena and Bulkley valleys run through the district and, with branches to the various mining groups, furnish easy access to the railway. An aerial tram is used by the Rocher De Boule Company from their property on Rocher Déboulé mountain to the railway, a distance of 4 miles with a difference in altitude of over 5,000 feet, whereas the distance by wagon road is 11 miles. Other properties on the mountain which depend on roads or trails are considerably handicapped in their development by the rugged character of the surface.

PHYSIOGRAPHY.

This district offers a striking contrast between the broad valleys of the Skeena and Bulkley rivers and the abrupt, rugged topography of the Rocher Déboulé group of mountains; this is well shown in Plate I.

The district lies immediately to the east of the Coast range of mountains and is a continuation of the Interior Plateau country of southern British Columbia; no indication of the plateau feature extends this far north, but the country presents isolated groups of hills and mountains, of various altitudes, some of which are subdued in type, whereas others are extremely rugged; these groups are separated by broad valleys which contain tumultuous streams, all of which are actively eroding and many of which flow through canyons.

The main valleys have all been eroded in rocks which are very much softer than those which form the hills; further, the cores of the hills are of igneous rocks in small stocks and these have metamorphosed and hardened the surrounding rocks, thus forming hard, resistant areas in a generally soft country. The rivers worked their way around these areas, choosing the paths of least resistance, ultimately forming the broad valleys and leaving the isolated hills.

GLACIATION.

Valley glaciation has accentuated the topographic contrast | truncating the spurs of the ridges and by filling up minor irregularities with debr'. The glaciers attained an elevation of over 5,500 feet and created, or accentuated, a subdued topography up to that height (See Plate IIA). Above the limits of the main glaciation is found a very rugged and serrated topography produced by recent glaciation, and many of the small local glaciers are still in existence, though they do not show much sign of activity. In the Rocher Déboulé group this type of topography is exceptionally well developed since the granodiorite lends itself to the formation of sharp pinnacled arêtes.

57091—2

The boulder clay in this district has been reworked in part by the rivers, and partly removed; some of it has been redeposited as finely stratified muds and sands. Samples of these muds from the Bulkley River valley near Smithers have been tested and a report upon them is given on page 00. The plastic character of the boulder clay cut through and used as filling material for the Grand Trunk Pacific railway is the cause of the slides which have occurred along the line.

Glacial striæ are found on the west side of Glen mountain in three sets—one north 13 degrees east, one north 21 degrees east, and an older set north 6 degrees east overridden by the other two. On Ninemile mountain, at 5,000 feet altitude, towards the west side, the striæ are north 31 degrees east, and the movement was apparently north to south. All the above bearings are with reference to true north. In most cases the hills were covered with vegetation, or were weathered so that no striæ were visible.

Besides the two main rivers, the Bulkley and the Skeena, which traverse this district, there are numerous smaller streams tributary to them, which reach all parts of the area. All these streams, especially those radiating from the Rocher Déboulé group, have very steep gradients and would furnish considerable water-power (See Plate IIA). The Bulkley is a turbulent river, flowing through a deep, narrow canyon for several miles above Hazelton and is capable of furnishing all the power necessary for the whole district.

GENERAL GEOLOGY.

The bedded rocks of the Hazelton district are all of the Hazelton series, described by Leach from the Telkwa district[1] and extending north into the Groundhog, as noted by Malloch[2]. Extending across the southern part of the Rocher Déboulé group of mountains, are interbedded flows and coarse, ill-assorted tuffs or tuff-agglomerates (See Plate IIB). North of the central portion of the group the series becomes more and more evenly-bedded, with well-assorted material, distinctly banded, and with very slight gradation from one band to the next, but all are tuffaceous in composition; this is well illustrated by Plate IIIA.

The beds exposed in the canyon of Bulkley river contain an abundance of plant fossils in some layers, and similar occurrences of fossils are found northward to the limits of the sheet, but much more sparingly. W. J. Wilson reported that these fossils seem to indicate the very top of the Jurassic or perhaps the lowest Kootenay rocks. Near the northern edge of the sheet, on Ninemile mountain, there is an horizon containing an abundant marine fauna mostly of pelecypods. A determination by Dr. T W. Stanton of the marine fossils collected, places the age of this series as "most probably upper Jurassic."

A folding of these rocks, which is well seen in the Bulkley canyon above New Hazelton, took place with axes approximately northeast-southwest (true). This folding was followed by the intrusion of small stocks or batholiths which now form the cores of Ninemile, Fourmile, and Rocher

[1] Geol. Surv., Can., Sum. Rept., 1910, p. 91.
[2] Geol. Surv., Can., Sum. Rept., 1912, p. 76.

Déboulé mountains, and by dykes which traverse both the mountains and the valleys (*See* Plate IIIB). The intrusions apparently cut through the sediments without causing any appreciable deflexion of the bedding.

<h2 style="text-align:center">PALÆONTOLOGY.</h2>

A collection of fossils was made from a marine horizon, in the Hazelton series, which passes through the American Boy property and is exposed by the underground workings there. This collection was sent to Dr. T. W. Stanton, of the United States Geological Survey, who reported as follows:

"The two lots of fossils from near Haselton, British Columbia, contain only a few recognisable forms.... The most abundant form in both lots is a large smooth Pecten which is very closely related to an unnamed Pecten in the *Aucella*-bearing upper Jurassic of southwest Alaska near Herendeen bay. In one lot there is also a single specimen doubtfully referred to *Aucella*, some poorly preserved specimens that seem to be *Ostrea*, and one or two other undetermined pelecypods. In the other lot, in addition to the *Pecten* and an *Ostrea*, represented by several imperfect specimens, there is one shell which appears to belong to the genus *Pedalion*, more commonly called *Perna*. While the fauna lacks strictly diagnostic forms I believe it to be most probably upper Jurassic."

A great many of the sedimentary tuffs of the Hazelton series, in this district, contain fossil flora; some of the beds contain an abundance of the fossils and collections were made from several horizons in the Bulkley canyon between the low and high level bridges and at three places on the road to Ninemile mountain. W. J. Wilson examined these collections and summed up his determinations as follows:

"In this collection there are a large number of specimens but only a few identifiable species. The most numerous and best preserved are those I have named *Cladophlebis virginiensis* Fontaine, *Ginkgo digitata* Brongn, and *Czekanowskia*.

"There are some good fronds of *Cladophlebis* which seem to agree specifically with *Cladophlebis virginiensis*, while others show considerable variation, and might be easily placed in some other of the many species into which the genus has been divided. Unfortunately the venation in most specimens is rather indistinct.

"Those named *Ginkgo digitata* may very well belong to the variety with much divided, narrow segments, and similar specimens are so placed by Knowlton and Seward. There seems little doubt that they belong to this species.

"At nearly all the localities there are specimens more or less covered with fine, narrow, needle-like leaves, either in tufts or single. These probably belong to the genus *Czekanowskia* and may be *C. murrayana* (Linley and Hutton) (= *C. rigida* Heer). There are also a number of tips of a delicate fern which may be a *Thyrsopteris*. This name Berry has changed to *Onychiopsis*. What may be a fragment of a stem of *Equisetites* is on 392-8 and a probable fruit is on 393-11. On 396-1 there is a *Ginkgo*-like form with narrow segments, which may be placed in the genus *Baiera*. There are good specimens of a cycad on one slab, which seems to agree best with *Dioonites borealis* Dawson; this specimen was an erratic picked up on the trail but supposed to be not far removed from the place of deposition.

"This meagre flora seems to indicate the very top of the Jurassic or perhaps the lowest Kootenay rocks."

It should be noted here that the sedimentary part of the Hazelton series in this district rests upon a series of crystal and of agglomeratic tuffs; the contact passes between the Cap and the Haselton View properties, on Rocher Déboulé mountain, and extends over the ridge about one-half mile south of the Rocher De Boule tramway. There appears to be a structural unconformity, but this was not conclusively demonstrated, however, there is a conglomerate at the base of the sediments, which contains pebbles and boulders of the underlying tuffs. This conglomerate is well exposed on the ridge above mentioned.

Another massive conglomerate occurs apparently in the middle of the sedimentary series and is well exposed in the canyon of the Bulkley river just below the high-level bridge; there does not seem to be any break in the series at this point, and the fossils are not sufficiently diagnostic to determine any slight break if one does occur.

ECONOMIC GEOLOGY.

GENERAL STATEMENT.

The district is divided into two parts by Bulkley river. The southern or Rocher Déboulé area is characterized by deposits of chalcopyrite carrying some gold and silver; molybdenite is important in one property and wolframite occurs in appreciable amount in one pros ect. The northern portion of the district includes Ninemile, Fourmile, and Glen mountains, and is characterised by silver-lead deposits.

South of Bulkley river the Rocher De Boule mine is the only large producer at present, but development work is being pushed on the Delta, Haselton View, Cap, and Golden Wonder properties, from all of which small shipments have been made. There are several properties on which a considerat. amount of work has been done, but they were shut down because of lack of capital, or because the ore opened up was of too low a grade, on the average, to permit of economical mining without milling facilities.

In all of the above-named properties, except the Hazelton View, the principal mineral is chalcopyrite and the ore carries small values in silver and gold; in the Haselton View property molybdenite and gold are both important and there is practically no chalcopyrite in the main vein. Wolframite occurs in important amount in a prospect on the Black Prince group, but has not been reported from any other property in the district.

The ore occurs in distinct shoots in veins of true-fissure-replacement, or shear-zone-replacement types, and in some cases one type passes into the other along the strike.

A scheme of general co-operation among the various companies that have proved their properties or have showings sufficient to warrant further development would be a great boon to the whole district. The lack of power, mill facilities, and means of transportation is felt by all the properties iper Creek basin except the Rocher De Boule, and to install a separate p nt for each property would involve an expense that would be prohibitive.

North of Bulkley river the Silver Standard mine is now the only large producer in the silver-lead group, but the American Boy is again opening and will ship this year. The ore in both these properties is in fissure veins in tuffaceous sediments and neither property is near the contact of a large intrusive. The veins are characterized by galena, sphalerite, and tetra-bedrite in a siliceous gangue. The tetrahedrite approaches freibergite in composition, for it carries over 2,000 ounces of silver in some places and is in sufficient amount to bring up the general silver values of the ore to a noteworthy extent.

The ore on these properties occurs in distinct shoots and the high grade material is associated with so n lower grade, or milling ore, that a custom mill has been erected by the Silver Standard. The output can thus be greatly increased from the Silver Standard mine and, by treating the ore from the American Boy, that property also will become a producer, since the high grade ore from the American Boy is not sufficiently bunched to permit of separate mining.

Near the contacts of the igneous stocks of Ninemile and Fourmile mountains the silver-lead deposits are characterised by a considerable percentage of jamesonite in addition to the minerals present in the two properties described above. None of the numerous properties in this class is at present producing, but some of them could become so if they had suitable milling facilities. The properties of Ninemile show much more promise than any seen on Fourmile mountain. The former have strong veins which can be traced for considerable distances along the surface, and which have very promising showings, with good silver values. The Shegunia river, at the base of the mountain, is about a mile distant from the properties and would afford a good site for a mill and for the develop-ment of more than the necessary water-power for the mines. An alternative plan would be to take the ore by aerial tram 6 miles south to a mill-site on Bulkley river, right at the railway.

MINING PROPERTIES SOUTH OF BULKLEY RIVER.

Properties in Juniper Creek Basin, Rocher Déboulé Mountains.

ROCHER DE BOULE MINE. *General Statement.* The Rocher De Boule mine is the most important producer in the northeastern mineral survey district of British Columbia. It has been shipping since April, 1915, has installed a good plant and transportation system, and has paid off all liabilities besides making considerable returns on the investment. The development work consists of more than 2 miles of main crosscuts and drifts, 2,200 feet of raises, and 330 feet of winzes.

The production to date has been 5,800,000 pounds of copper, 51,340 ounces of silver, and 3,736 ounces of gold. About 90 per cent of the copper was taken from high grade shoots above the 500-foot level on No. 4 vein, the highest, and the remainder from a high grade shoot on No. 2 vein between the 1,000-foot and 1,200-foot levels, which is still being worked.

In opening up the high grade ore much second grade or milling ore was encountered and the company intends to install a mill to handle this product, the returns from which are expected to carry the cost of further search for high grade shoots.

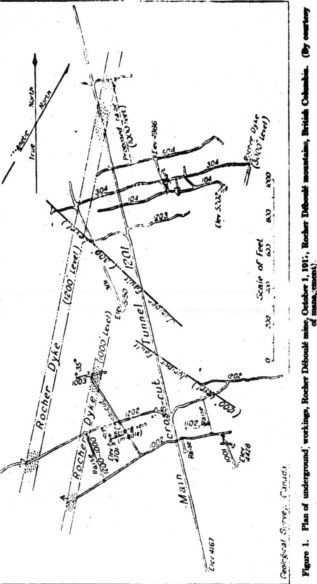

Figure 1. Plan of underground workings, Rocher Déboulé mine, October 1, 1917, Rocher Déboulé mountain, British Columbia. (By courtesy of management).

A crosscut tunnel has been run into the mountain for more than 3,100 feet at the 1,200-foot level and intersects four veins, giving a back of 180 feet for the lowest vein, 460 feet for the next higher, 1,050 feet for the tramway vein, and 980 feet up to the 500-foot level of No. 4 or the main vein. The work at present consists of drifting on the lowest level of No. 4 and of raising .a it to the 500-foot level, with intermediate drifts to explore the vein between the levels; at the same time development is being pushed on No. 2 vein and good ore is being encountered, including a shoot which is more than 70 feet long and averages 2 feet in width running 7 per cent copper.

The Rocher De Boule property is located on Rocher Déboulé mountain (Map 1732) on the west side of Juniper creek 1¾ miles from its head, and extends from the creek at 3,900 feet elevation, to the top of the ridge at 5,700 feet elevation. A series of approximately parallel fissures traverse the hillside with a general strike north 60 degrees east (magnetic) and with dips averaging about 60 degrees northwest. These fissures have been formed by shearing, and the crushed country rock has been partly replaced by mineralizing solutions, so that the fissures have become veins containing valuable deposits of ore.

The veins are not of equal economic importance, nor are all parts of the same vein equally valuable (Figure 1); and, although all the veins are mineralized, only two of the five have been shown to contain high grade ore-shoots in addition to the second grade or milling ore which is found in greater or less amount in all the veins. In the upper part of the highest vein, for instance (Figure 2), there were four large bodies of high grade copper ore of irregular shapes at approximately the same elevation, but separated horizontally by 50 to 200 feet of vein material carrying much lower values, and, in places, no values at all.

Factors Influencing Location of Ore-shoots. From the data at hand the following statements may be made.

The wall-rock was not an important factor in determining the position of the ore-shoots, if indeed it had any influence. This is inferred because there is apparently no difference in its composition near the shoots from that bordering barren portions of the veins.

Cross fissures are not common. In places, slightly inclined fissures are filled with ore near the main shoot; but the condition is not so general as to indicate that these fissures were a determining factor in the location of the shoot.

Dykes parallel some of the veins, now on one wall, now on the other, and are cut by them. One large cross dyke has been recently shown to cut through the property, but whether or not it has influence on the ore deposition has not been determined (*See* Figure 1, Rocher dyke).

The gangue does not appear to have exerted a deciding influence, for the ore is found in varying amounts in all the gangues apparently without discrimination, except that actinolite is found in all the important shoots, mixed with the apparently solid sulphides and more or less replaced by them.

Two major periods of fissuring, parallel to the veins and expressed in the veins, have been demonstrated and they have an important bearing on the location of the ore-shoots.

The primary fissuring was followed by an alteration of the brecciated and ground-up country rock along the fissures, together with a consider-

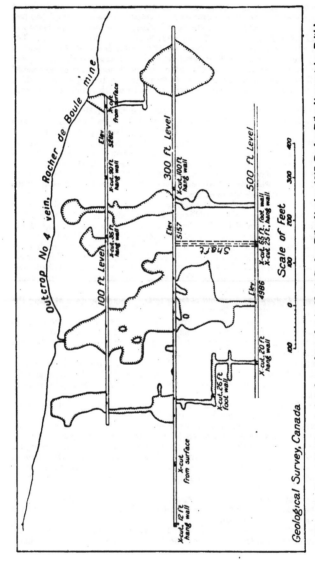

Figure 2. Stope elevations projected on plane of vein No. 4, Rocher Déboulé mine, 1917, Rocher Déboulé mountains, British Columbia. (By courtesy of management).

able development of actinolite in places; a general silicification bound the whole mass together again, and filled the interstices with quartz. The result then was veins, with fairly definite walls which were parallel in general but not in detail, which bulged and contracted both horizontally and vertically to include the brecciated and partly ground-up country rock (granodiorite), or followed relatively narrow and definite lines where the rock was not shattered. The filling in the wider parts would then be of fragments and debris of the original rock, partly or in places wholly replaced by actinolite and quartz in varying amount, whereas the narrower parts contained similar material but no fragments, so were much tighter.

The second fissuring followed the same lines as the primary, but in this case the fissured rock was not homogeneous as at first, so that the fissure had a different expression in passing through the different gangue materials or combinations of them. In the strongly cemented, narrow portions the fracturing seems to have made a relatively clean break either in the vein or along one of its walls, showing that the vein in those portions acted as though it were part of the general country rock. In the heterogeneous portions of the vein the fissuring spread, giving the effect of crushing.

The ore-bearing solutions followed the second fissuring, and naturally found easy passage through those parts of the veins which had been crushed most by this fissuring; these places were also the loci of the greatest deposition of actinolite and of crushed and altered country rock, both of which were apparently easily replaced by the ores.

The primary cause of the larger portions of the vein is not evident. They may have been partly due to differential movement of the irregular walls of the fissure, otherwise it is not clear why certain portions of the granodiorite should have suffered brecciation and why others, apparently similar, broke with a clean fracture. This point seems to be the ultimate determining factor in the location of the ore-shoots.

Paragenesis of the Ores. General. There were two distinct mineralizations in the veins on this property. Their sequence can be established, but the lapse of time between the depositions is not readily determinable. The first mineral group bears a close relationship to that of the Delta property, but is distinctive in having a considerable development of arsenopyrite and of pyrrhotite instead of the abundant magnetite and hematite of the Delta occurrence. The other minerals are the same in each.

In position the Rocher De Boule veins are nearer the contact of the granodiorite with the sediments than are the Delta veins and apparently represent a progressively cooler phase of the deposition, the hottest portion of which would be expected in the interior of the granitic mass where the Delta property is located.

The ore from the veins of the Rocher De Boule mine is chiefly valuable for its copper content, though paying quantities of silver and gold are also present. Besides these metals, large amounts of iron, lead, zinc, and arsenic, and smaller quantities of antimony and molybdenum also occur, but none of them is worth extracting

These metals are combined in the following minerals: chalcopyrite, which forms the bulk of the ore; pyrite, pyrrhotite, arsenopyrite, tetrabedrite, zinc blende, galena, which may be seen in almost all hand specimens, and magnetite and molybdenite which are only rarely seen.

57091—3

The gangue minerals associated with these are hornblende, glassy quartz, banded milky quartz, and small amounts of calcite, siderite, and tourmaline. The hornblende greatly exceeds in amount all of the other gangue minerals. Tourmaline was found in only one or two places and in quite small amounts.

Two types of ore are found in this mine which differ markedly in appearance and mineral composition, and which usually occur in different parts of the mine, though in one or two places the two kinds were found in contact. The more plentiful of the two types and that which produces the bulk of the copper may be called the chalcopyrite-hornblende ore, for these two minerals constitute fully 90 per cent of the vein material. The chalcopyrite is usually in great excess over the hornblende, and consequently this is an unusually rich copper ore. The other type consists of large quantities of banded milky quartz with relatively small quantities of zinc blende, tetrahedrite, galena, and chalcopyrite. The silver content is more valuable than the copper and this ore will, therefore, be referred to as the silver-lead ore. It is largely confined to No. 2 vein and is best developed in the 1202 drift to the west of the main tunnel.

Chalcopyrite-hornblende Ore. This ore contains, besides the chalcopyrite and hornblende, a considerable amount of pyrrhotite, arsenopyrite, pyrite, magnetite, tetrahedrite, and molybdenite. A fairly large amount of quartz occurs in some of the ore-shoots, but it is distinguished from the quartz of the silver-lead ore by having no banding, and by being clean and glassy instead of milky. The tourmaline is very subordinate in amount and may be considered only as a rare gangue mineral in this deposit. Calcite and siderite are much more common. The group of minerals as above enumerated strongly suggests high temperature deposition.

The chalcopyrite presents nothing unusual in its appearance either in hand specimen or under the microscope, and large masses of it seem to be quite homogeneous; however, Mr. J. D. Galloway, of the British Columbia Bureau of Mines, claims that analyses show it to vary considerably from the theoretical composition $CuFeS_2$. It is associated more closely with hornblende than with any other mineral, and was frequently seen replacing crystals of hornblende in a manner similar to that shown in Plate IVA. On the other hand numerous places were observed in the polished sections where fringes of actinolite followed the contacts of the chalcopyrite, and feathered out in a manner which clearly shows it to have been deposited later than the chalcopyrite. This is shown in Plate IVB and it will be observed that this actinolite is also later than the pyrrhotite and magnetite. The chalcopyrite was seen to have replaced magnetite, pyrrhotite, pyrite, arsenopyrite, molybdenite, and quartz. It was found to be replaced by molybdenite and, in the proximity of the silver-lead ore, by arsenopyrite, tetrahedrite, and galena. Occasionally tetrahedrite is veined by chalcopyrite (See Plate VA). The replacement by arsenopyrite is shown in Plate VB. The chalcopyrite which forms a large shoot in vein No. 2 is fractured and sheared, the cracks being filled with calcite, and in the same specimen calcite was observed which was clearly replaced by chalcopyrite.

Pyrrhotite was found in all the big chalcopyrite ore-shoots, but seemed to be much more plentiful in vein No. 4 than in No. 2, though it was observed in the latter in the 1002 stope. It occurs as irregular masses in the chalcopyrite, and some evidence was found to show that it was deposited earlier than the chalcopyrite, but this was not conclusive. This mineral does not show any cracks nor does it contain veinlets of other minerals, and, since such structures are the surest indicators of paragenesis, it is often difficult to tell the relative age of the pyrrhotite. It is considered by most economic geologists to be almost exclusively a high temperature mineral, and though the evidence is not conclusive that the chalcopyrite is later than the pyrrhotite still there is nothing to suggest that the pyrrhotite is the later; therefore, it may safely be considered that the natural sequence of deposition of pyrrhotite-chalcopyrite was followed.

The pyrrhotite is seen replacing magnetite grains and the magnetite is apparently the only mineral which preceded the precipitation of the pyrrhotite.

Arsenopyrite is not plentiful in the Rocher De Boule veins, but is of two different periods. In No. 4 vein it is undoubtedly earlier than the chalcopyrite, and numerous examples of etched and veined crystals were seen. In No. 2 vein a reversed order is found in which beautiful, small veinlets of arsenopyrite traverse the chalcopyrite or follow its gangue contact in a most regular and persistent way. This is well illustrated in Plate VB. This specimen, from which the photograph was taken, is from a point in the vein where the silver-lead ore comes in contact with the chalcopyrite-hornblende ore, and it is believed that the arsenopyrite shown in it belongs to the same late period of mineralization that produced the silver-lead ore, whereas the chalcopyrite belongs to the earlier period in which the chalcopyrite-hornblende ore was deposited. The arsenopyrite was found veining hornblende and is, therefore, of later deposition. No safflorite, gold, or löllingite, were observed in any of this arsenopyrite, though they are plentifully associated with the arsenopyrite of the veins on the Hazelton View claims, a short distance to the west.

Pyrite is only sparingly present and its structures indicate that it was deposited shortly after the magnetite, and probably about the same time as the pyrrhotite. It occurs as small disseminated crystals and is frequently veined by chalcopyrite.

Molybdenite is the least plentiful of all the minerals present and is seen only as small flakes sparsely disseminated through the hornblende. Its relative time of deposition is doubtful, but it was seen replacing arsenopyrite and chalcopyrite, and was probably deposited soon after those minerals (See Plate VIA).

Tetrahedrite occurs in this ore, but in much smaller quantities than in the silver-lead ore. It seems to have been deposited with the chalcopyrite or slightly preceding it.

Much later than all the minerals so far described, and following a period of considerable crushing, small veinlets of calcite and siderite were formed.

Silver-lead Ore. This ore consists of large masses of banded milky quartz containing a small proportion of zinc blende, galena, tetrahedrite, pyrite, arsenopyrite, and chalcopyrite.

Pyrite and arsenopyrite are present only in minute amounts and are extensively replaced by tetrahedrite and zinc blende. Tetrahedrite is the most abundant ore mineral in this type of ore; it is frequently seen sprinkled throughout the ore, sometimes quite plentifully, with small etched crystals of arsenopyrite, and associated with each there is usually a small area of chalcopyrite. This structure is so common in the tetrahedrite that it might be said to be characteristic of the tetrahedrite of this ore. The amount of chalcopyrite in the ore is very small, and little can be seen besides the small blebs mentioned as occurring in the tetrahedrite. Galena is more abundant but does not equal in amount the tetrahedrite or zinc blende.

The paragenesis is clearly shown to be as follows:

Table of Paragenesis of the Ores of the Rocher De Boule Mine.

First mineralization
Granodiorite
Fissuring
Sericite and chlorite formed in crushed zones
Development of actinolite
Silicification
Second fissuring
Magnetite
Pyrite and arsenopyrite
Molybdenite
Pyrrhotite and chalcopyrite
Second mineralization
Arsenopyrite and pyrite
Zinc blende }probably sometimes contemporaneous
Tetrahedrite}
Chalcopyrite }often reversed
Galena }
Last phase
Development of siderite and veining by calcite
Products of oxidation

Actinolite is an important gangue mineral which in some sections appeared to come in after the silicification and before the crushing, in others later than the first chalcopyrite.

The conspicuous absence of pyrrhotite, molybdenite, and hornblende in the second mineralization would seem to indicate that the temperature of deposition of the second was much lower than that of the first mineralization.

DELTA PROPERTY. *General Statement.* The property of the Delta Copper Company immediately adjoins that of the Rocher De Boule Company (*See* Map 1732) and extends for about 3,500 feet east. It is on the north side of Juniper creek, and extends from the creek bed up to over 7,000 feet in altitude.

Description of Main Veins. General. There are a number of veins on this property, but only two have received much attention up to the present. The lower of these two is thought to be a continuation of the fissuring on the upper part of the Rocher De Boule property, of which No. 4 vein is the main expression. It outcrops at intervals along the hillside between 5,500 feet and 6,000 feet elevation, and has been traced from one end of the property to the other. The strike of this vein is north 85 degrees west, and its dip is 70 degrees to the north. The upper vein outcrops at intervals across the property between elevations 6,200 and 6,700 feet. It strikes north 78 degrees west and dips 75 degrees north.

The Lower Vein. The lower vein varies in width from 2 to 3 feet at the east end of the property where a short tunnel is located at 5,900 feet elevation. A sample taken from across 20 inches gave the following values: gold, trace; silver, 0·5 ounce per ton; copper, 1·5 per cent. Two hundred feet west, at the McDonell cut (*See* Figure 3), the vein is 4 feet wide with 2 feet of low grade ore. Two thousand feet to the west of the tunnel, stripping shows a vein 8 feet in width with 2½ feet of magnetite, hematite mostly, and some chalcopyrite, assaying—gold, 0·01 ounce; silver, trace; copper, 1·10 per cent. The remaining 5·5 feet is of decomposed material which assayed: gold, 0·01 ounce; silver, 0·40 ounce; copper, 0·14 per cent. A very fine exposure of ore was opened near the last-mentioned stripping,

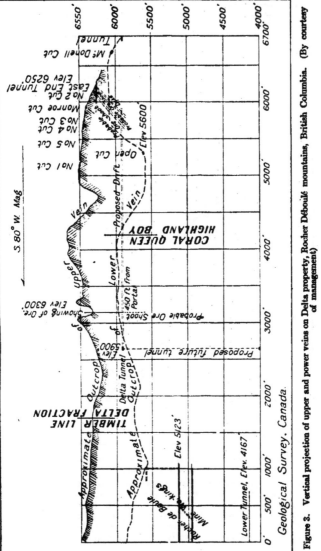

Figure 3. Vertical projection of upper and power veins on Delta property, Rocher Déboulé mountains, British Columbia. (By courtesy of management)

during the summer of 1917, showing 2 feet of high grade copper ore in the vein which is here 12 feet wide.

The Upper Vein. The upper vein has been traced and partly stripped for a distance of 3,500 feet along the strike. The top part is usually highly oxidised and leached to a depth of from 3 to 10 feet from the surface, but

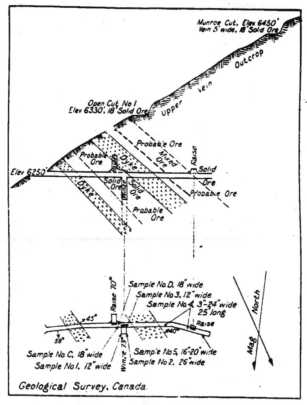

Figure 4. Plan and elevation of east end workings, Delta property.
(By courtesy of management.)

has been cut through in a number of places to expose the fresh material below. The vein at the most westerly exposure is at an elevation of 6,300 feet; at this place three open-cuts and strippings have exposed an ore-shoot practically continuous for 200 feet along the vein and from 1 to 6 feet in width. Although many parts of this shoot show oxidation and leaching, other parts are practically unaltered and a sample across 18 inches of the latter gave 20 per cent of copper; and in the main cut a sample across 6 feet gave 6 per cent of copper (assays furnished by Mr. Clement).

From this place towards the east, open-cuts show a strong vein with distinct walls, which varies in width from 3 to 15 feet and averages about 5 feet. Only in a few places has the oxidized portion been completely cut through, notably near the east end.

At the Munro cut, 350 feet from the east end tunnel, the vein is 5½ feet in width and shows 18 inches of solid chalcopyrite in the middle; the rest of the vein is highly oxidized and leached. At No. 2 cut, 150 feet west of the tunnel, the vein is 4 feet in width; there are 18 inches of high grade ore on the hanging-wall and the remainder is of low grade ore.

The tunnel at the east end of the Delta property (Figure 4) on this vein enters the vein at an elevation of 6,250 feet and runs for over 200 feet, thus bringing it well under No. 2 cut, giving 115 feet of backs at this place, for a shoot 27 feet long and 18 inches wide of high grade ore. The tunnel develops no ore for the first 65 feet from the portal, then 7 inches of solid sulphide appears in the vein which at this place is 3 feet in width, and is continuous for the next 27 feet, varying from 7 to 18 inches in width and following the hanging-wall; besides the high grade there is generally 18 inches to 2 feet of low grade ore along the foot-wall. This ore ceases abruptly where the vein cuts through a dyke at this place for 33 feet. A few feet from this dyke the ore was again encountered, varying from 3 to 24 inches in width in a distance of 35 feet, but of lower grade than before. For the next 27 feet there is a band of solid ore on the hanging-wall from 12 inches to 24 inches in width, which is probably the downward extension of the ore exposed in No. 2 cut.

At 85 feet from the portal a raise was put up in the ore, but at 15 feet above the drift the ore stopped when the dyke was encountered. A winze was sunk a few feet west of the raise to a depth of 30 feet, in 12 inches of ore throughout this depth; at the bottom of the winze the ore was 12 inches wide at the east side and 26 inches wide at the west side, all of high grade.

Paragenesis of the Ores.

Table of Paragenesis of the Ores from the Highland Boy Veins, Delta Property.

Granodiorite.
Fissuring, with development of sericite and some chlorite, with some silicification of the adjacent country rock.
Actinolite replacing altered country rock in places along the fissures.
Silicification and partial replacement of altered country rock and of the actinolite by quartz.
Crushing along the same lines as before, showing especially in the more siliceous portions.
Vein replacement by metallics in the following order:
    Magnetite.
    Hematite.
    Molybdenite (in small amount and position approximate).
    Pyrite.
    Gold (a few grains replacing magnetite and quartz came with pyrite and chalcopyrite).
    Chalcopyrite, with a little tetrahedrite and bornite.
    Tin mineral (HD_2O) trace, paragenesis not known.
    Veining and replacement by calcite.
It should be noted that none of the lead-zinc minerals is present in the veins on this property.

Discussion. In general the history of the veins on the Delta property has been similar to that of the Rocher De Boule veins; the main points of difference are in the relative proportions of the metallic minerals present and in the fact that the lead-zinc mineralization is not present in the Delta veins.

18

Magnetite is the most abundant of the metallic minerals and it is accompanied by a small amount of specularite. These were the first of the metallics to be deposited, apparently a little before the molybdenite which is small in amount. No pyrrhotite or arsenopyrite were observed, but pyrite followed the molybdenite in considerable amount. Chalcopyrite is next in abundance to the magnetite and is usually associated with a little tetrahedrite and bornite; it replaces the gangue, magnetite, and pyrite principally. Small round grains of gold are seen in the magnetite, bearing the same relation to it as does the chalcopyrite; otherwise the paragenesis could not be proved. Although no ore of tin was distinguished under the microscope, chemical analysis shows that there are traces of tin in the ore. On the whole, this property holds considerable promise and under the energetic policy of development which is projected, it should soon become an important producer.

GREAT OHIO PROPERTY. *General Description.* The main vein on this property is approximately 4,000 feet south of the main vein on the Rocher De Boule, and is roughly parallel to it. The strike of the vein parallels the contact of the batholith with the tuffs about 50 feet inside the granodiorite at the tunnel, but farther up the hill passes into the tuffs where they cap the batholith. There are a series of parallel fissures on the property, but only the main one, 3 to 4 feet in width, is worthy of note.

The veins often follow small dykes of camptonite, cutting through them or on either wall; and the dykes themselves are somewhat mineralized. The chief gangues are quartz and actinolite.

The ore is not evenly distributed, nor does it occur in large, well-defined shoots, but rather in irregular bunches, or small streaks; one of these streaks, 4 inches across, gave an assay of gold, 0·04 ounce; silver, 134 ounces; copper, 1 per cent; and the total width of the vein at this point was 4 feet.

Table of Paragenesis of the Ores from the Great Ohio Property.

Fissuring of country rock
Dykes of camptonite
Fissuring following dykes frequently
Sericitization and silicification of country rock (granodiorite) with deposition of some vein quartz
Formation of actinolite in the veins
Arsenopyrite, pyrite, pyrrhotite
Marcasite ⎱ Not positive of position, may be later than galena-sphalerite group
Siderite   ⎰
Zinc blende
Tetrahedrite ⎫
Tennantite ⎬ Intimate mixture
Chalcopyrite ⎭
Galena
Crushing
Calcite and quartz
The peculiar occurrence of marcasite and siderite is shown in Plate VIB.

RED ROSE PROPERTY. *General Description.* There is one main vein on this property which strikes in a direction nearly at right angles to the veins on the Highland Boy claims. Although the vein is nominally on the sedimentary side of the contact, the granodiorite is so near the surface that portions of it protrude through the sediments, so that the vein is sometimes in the granodiorite, sometimes in the sediments.

The vein strikes up the hillside and is exposed by cuts and sluicing at intervals from 5,400 feet up to 6,500 feet elevation. The matrix in the lower portions is of quartz and ground country rock, but farther up the hill the filling is almost wholly of milky quarts. The vein is about 2½ feet wide in the lower part, widening to over 12 feet where it passes over a shoulder and under slide rock about 6,500 feet elevation; it evidently swings decidedly towards the west in the upper part.

The parts of the main batholith which protrude through the tuffs on this property contain very little quarts and the feldspars indicate that the rock is diorite or quarts-diorite rather than granodiorite.

The feature of this deposit is the development of considerable biotite and of tourmaline in connexion with the mineralisation.

Table of Paragenesis of the Ores from the Red Rose Property.

Fissuring, with some movement
Development of biotite, tourmaline, and milky quarts
Pyrite and arsenopyrite, latter containing safflorite
Pyrrhotite
Chalcopyrite
Marcasite, replaced by siderite and limonite
Covellite and native copper

The Red Rose vein is prospected by three adit tunnels: one at an elevation of 5,734 feet, 160 feet in length; one at 5,456 feet, 225 feet in length; and a third which was filled up with slide rock at the time of examination. Besides these, a crosscut tunnel was driven from an elevation of 5,202 feet to intersect the vein, but it was abandoned after going 400 feet without striking ore.

The mineralization is usually on the foot-wall of the vein which strikes north 70 degrees west (magnetic) and dips 50 degrees to the south-west. At the main tunnel the vein is 4½ feet in width and the paystreak 30 inches wide on the foot-wall. A sample across this ore gave: gold, 0·84 ounce; silver, 3·2 ounces; copper, 3·9 per cent; and across the remaining 2 feet of siliceous gangue the values were: gold, 0·02 ounce; silver, 1·4 ounces; copper, 2·1 per cent[1]. About 300 feet up the hillside from where the above sample was taken, another sample taken from across 18 inches gave: gold, 0·30 ounce; silver, 2·3 ounces; copper, 8 per cent[1].

This property is not producing, as enough shipping ore to prove attractive as an investment for quick returns has not yet been opened up, but it is kept in good condition and with facilities for having the ore milled within a reasonable distance of the mine it could produce at a profit and at the same time pay for the search for the high grade shoots.

BRIAN BORU PROPERTY. *General Description.* This property is situated just south of the area described in the present report, but it represents a type of deposit not prominent at other places on the Rocher Déboulé mountain, since it is altogether a zinc-lead proposition.

About 7 miles up from Skeena Crossing a good trail branches from the main Rocher Déboulé road and follows Brian Boru creek eastward to its head, a distance of about 4½ miles with a rise of 3,000 feet to the Brian Boru properties.

The claims extend across a spur of stratified, fine-grained, and coarse tuffs with some flows which are highly pyritised and are at this place about one-half mile west of the contact of the granodiorite.

On the north side of the spur a vein was traced for several hundred feet, striking north 50 degrees east with dips from 10 to 50 degrees to the northwest. The vein averages 10 feet in width, but varies from 4 to 15 feet, splitting and rejoining.

[1]British Columbia Bureau of Mines, 1915.

The vein filling is mostly calcite and country rock with a little quartz and hornblende, together with the metallics noted in the paragenesis. Besides the main vein there are stringers in the country rock containing galena and sphalerite.

On the south side of the spur there are two parallel veins; they strike north 5 degrees east and dip northwest 45 degrees. The first, 4 feet in width, is opened by short tunnels for several hundred feet, and is practically solid zinc blende cutting fine-grained, stratified tuffs. The second is opened by a single cut and exposes a mineralized zone, 12 feet in width, cutting coarse tuffs. This zone is made up of stringers of solid zinc blende cutting the country rock.

Table of Paragenesis of the Ores from the Brian Boru Property.

Actinolite and quartz
Pyrrhotite and pyrite (mostly earlier than the zinc blende)
Zinc blende (containing specks of contemporaneous pyrrhotite and chalcopyrite
Chalcopyrite (veining zinc blende)
Galena and jamesonite (veining chalcopyrite)
Calcite (in veins in all other minerals)

Zinc blende is the most abundant of the metallic minerals and the typical association of the minerals is shown in Plate VIIA. There is not much galena and even less chalcopyrite in the ore.

The Annual Report of the British Columbia Bureau of Mines for 1915 states that assays give little or no silver, gold, or copper.

A large vein containing mostly galena was reported to have been discovered and partly opened in 1917, but this showing was not seen during the examination.

The veins are large and strong and it is not clear why the silver-bearing tetrahedrite should be absent.

GENERAL STATEMENT. From the general conditions, this western side of the batholith should prove good prospecting ground, since part of the roof is preserved in places and it is in the sediments above the higher parts of the granodiorite or in the granodiorite immediately below the old cover that mineralisation is most probable.

Properties on the West Slope of Rocher Déboulé Mountains.

HAZELTON VIEW PROPERTY. *General Statement.* This property is owned by the New Hazelton Gold Cobalt Mines, Ltd., and adjoins the properties of the Rocher De Boule and Delta companies to the west (*See* Map 1731). The camp and workings are on the west face of the mountain so that direct access may be had to the railway.

The granodiorite-sediment contact crosses the centre of the property; some of the veins are wholly in the sediment and the main vein crosses from the granodiorite into the sediments with apparently little change.

The main tunnel is situated at about 5,500 feet elevation and is connected with the camp at 4,150 feet elevation by an aerial tram; from the camp to the railway the ore is being taken by pack train until a road is completed.

Several trial shipments have been made of the ore and development work is being continually pushed; two lower adit tunnels are being driven, and it is expected that this property will ship in 1918.

The ore is valuable for its content of gold and molybdenite especially, and contains little if any chalcopyrite.

Description of the Veins. There are several veins crossing the property of the New Hazelton Gold-Cobalt Mines, Ltd., of which the main vein is the most southerly. The veins strike north 80 degrees east (true) and the main vein dips 55 to 60 degrees to the northwest. Besides the main vein two others have been opened by cuts along the strike. One shows a crushed zone in the sedimentary tuffs about 4 feet in width and containing actinolite, quartz, and biotite, with some chalcopyrite; the other shows a 4 to 6-inch vein in the tuffs, containing galena and pyrite in a gangue of milky quartz.

The main vein contains gold-bearing arsenopyrite and molybdenite in a gangue of actinolite and quartz and strikes directly up the hillside, crosses the ridge, and descends the Juniper C eek side, where it passes under drift. The vein is closely associated with a 2-foot dyke of fine-grained, granodiorite-porphyry which parallels it either along one wall or a short distance from it. The width of the vein varies from 18 inches to 4 feet and it has been shown up by cuts and stripping for 2,200 feet horizontally thro gh an elevation of over 1,100 feet.

At 600 feet below the top of the ridge a drift tunnel has been driven on the vein for a distance of 625 feet. The first 85 feet in the tunnel discloses the vein 18 inches to 2 feet in width, carrying 4 to 18 inches of solid sulphides on the hanging-wall, which are said to average $80 in gold and 2½ to 5 per cent of cobalt. In stoning from this shoot, according to information given by Mr. Morkill, the first three tons gave an assay of 5·20 ounces of gold, 4 per cent cobalt, and 22 per cent MOS_2.

For the next 40 feet in the tunnel there is an open fissure instead of the vein; it varies in width from 3 inches at the ends to 2 feet in the middle, and extends to an unknown distance vertically. There is apparently not even a crustification of the walls of the fissure.

The open fissure terminates where the walls come together to form a vein 3 inches wide filled with gouge; the vein immediately widens again, but is barren of ore for 50 feet, when a small shoot is encountered. At 225 feet a dyke is cut through by the vein and has been offset for 11 feet, the north side having moved relatively towards the west. At 375 feet from the portal a raise was being put up in ore consisting of arsenopyrite and molybdenite in a gangue of actinolite, quartz, and calcite; a 9-inch streak here ran 20 per cent molybdenum and the general ore from the raise 3 to 4 per cent MOS_2. Hand-sorting gives a product carrying 7·4 per cent MOS_2 and $10 in gold.[1]

The gold content of the ore is very variable; for instance, one streak 4 inches across, just past the raise, is said to have assayed $631 in gold. It is generally considered in the mine that the finely crystalline arsenical iron carries about $18 to $20 in gold and when it is darker in colour and coarsely crystallized it carries from $50 to $100 in gold. Under the microscope there could be seen no reason for such a difference, as in both cases the so-called arsenopyrite is a mixture, as noted in the paragenesis; the high analysis quoted above was from fine-grained sulphides. Across 2 feet (the full width of the vein) of ore at the top of the ridge an assay shows: gold, 4·0 ounces; silver, 0·02 ounce; cobalt, 3·0 per cent.[2]

On the Juniper Creek side of the ridge an open-cut 6 feet deep exposes the vein, which is here 2·5 feet in width with 6 inches of sulphides and 2 feet of decomposed vein material. A sample across the whole vein gave: gold, 0·65 ounce; silver, 0·2 ounce; cobalt, 1·0 per cent.[3]

[1] Assays furnished by Mr. Dalby Morkill.
[2] British Columbia Bureau of Mines, 1917.
[3] British Columbia Bureau of Mines, 1917.

About 300 feet below the tunnel the vein passes from the granodiorite into the surrounding sediments. The vein continues strong and exhibits characteristics similar to those on the granitic side of the contact.

Analysis of Sample of Arsenical Ore, from the New Hazelton Gold and Cobalt Co., Ltd., New Hazelton, B.C.[1]

Silica and insoluble silicates	13·70	per cent
Iron	17·98	"
Arsenic	55·53	"
Cobalt	4·40	"
Sulphur	4·00	"
Lime	2·07	"
Magnesia	0·90	"
	98·58	

Gold, 4·36 ounces a ton = $87.20
Silver, 3·60 ounces a ton = 3.77
Specific gravity, 5·66

Paragenesis of Ores. The ore from the main vein on the Hazelton View property is mined for its molybdenum and gold values. It contains, besides these metals, a large amount of arsenic, iron, and cobalt, and some silver, which are combined in a group of minerals whose association is most unusual if not quite unique.

The gangue in which these minerals are found is similar to that of the Rocher De Boule veins, since it consists chiefly of hornblende with a small amount of quartz. Feldspar is also found in the vein, but it is thought to be supplied by the crushed wall rock.

The metallic minerals present, named in the order of their abundance, are: arsenopyrite, safflorite, löllingite, molybdenite, native gold, a little chalcopyrite, and probably some electrum or telluride carrying silver.

The arsenopyrite is usually coarsely crystalline and is replaced by safflorite, löllingite, molybdenite, and the gold and silver minerals with a little chalcopyrite. It is later than the hornblende and is often seen in veins cutting through the latter.

The safflorite is found chiefly in the arsenopyrite which it replaces, but a considerable amount also occurs as irregularly shaped grains in the hornblende. It differs from the arsenopyrite in having a slightly whiter colour, a smoother surface in the polished section, and a lower relief due to its slightly less hardness, also by its apparent inability to develop crystal faces against the gangue which it replaces. This last habit is in marked contrast with the arsenopyrite which invariably develops sharp, brilliant, crystal faces against the hornblende. The safflorite was determined in the polished sections chiefly by its behaviour on treatment with ferric chloride. This reagent etched it very slowly, developing a pronounced cleavage in one direction. This single, strong cleavage served to distinguish it from smaltite which is isometric, the safflorite being orthorhombic.

Löllingite is fairly plentiful and is closely associated with the arsenopyrite, though their relative ages were not determined. It is very similar to the arsenopyrite, but is distinguished by its comparative indifference toward nitric acid.

The molybdenite is fairly plentiful and much of the material mined forms good molybdenum ore.

The molybdenite is seldom found in the arsenopyrite, but is freely disseminated in the hornblende as small flakes and crystals. Its relation to the arsenopyrite seems to indicate that it is later in order of deposition.

The gold occurs as small grains scattered through the arsenopyrite and safflorite, and, excepting a very small amount which was seen in the quartz,

[1] Analysis by James G. Powell, Provincial Assayer.

it is confined to the arsenic minerals. Associated with the gold in grains similar in size and shape to the gold grains, occurs another mineral varying from yellowish cream to creamy white in colour, which is thought to be electrum, but as yet is not definitely identified. The gold in arsenopyrite and safflorite is shown 'n Plate VIIB. Some gold was found occupying minute veins in the a' .....nopyrite and was, therefore, deposited some time after the arsenopyrite.

Table of Paragenesis of the Ores from the Main Vein.

Granodiorite
Intrusion of dykes.
Faulting, with throw of 11 feet
Sericitisation and development of abundant actinolite
Silicification and replacement of other minerals by quartz
Magnetite
Pyrite
Arsenopyrite, safflorite, löllingite
Gold
Molybdenite
Calcite

The gold may have been contemporaneous with the arsenopyrite since it is always embedded in that mixture; the molybdenite comes later than the arsenopyrite mixture. The actinolite in several sections appears to be later than the molybdenite; it is probable that the development was in two stages or was continuous during the first mineralisation. There appears to have been some fissuring, with a deposition of silica at some time between the deposition of arsenopyrite and the deposition of molybdenite; this is well shown in Plate VIIIA.

PRESTON PROPERTY. *General Statement.* Going west from the Hazelton View workings, down the mountain, at elevation 2,850 feet, is the Huckleberry claim which contains a fissure 3½ to 4 feet wide striking north 80 degrees west (magnetic) and dipping 69 degrees to the southwest. A stripping exposes this vein for 150 feet through coarse tuffs. The filling is mostly crushed country rock with some pyrite and arsenopyrite, carrying a little silver and gold.

CAP PROPERTY. *General Description.* The Cap group is situated at 2,200 feet elevation, practically on the strike of the Hazelton View veins, distant 9,000 feet horizontally and 3,000 feet vertically. It has been opened up by strippings, by a 60-foot shaft, and by two short crosscut and drift tunnels. The vein is 18 inches to 5 feet in width and cuts crystal tuffs which have been mistaken for granodiorite. The gangue is mostly of crushed and altered wall rock with some quartz. The ore is valuable for its silver and copper, and carries a little gold.

At 20 feet down the shaft a sample across 2 feet of the vein gave: gold, 40 cents; silver, 9·8 ounces; copper, 1 per cent.[1]

At 10 feet in the shaft a sample across 2 feet of ore gave: gold, 0·04 ounce; silver, 21·4 ounces; copper, 7·5 per cent.[1]

A sample of 5 or 6 tons of the best ore from the upper 20 feet of the shaft gave: gold, $1.20; silver, 25·2 ounces; copper, 9·7 per cent.[1]

The first tunnel came in 40 feet below the collar of the shaft, and passed through a small shoot which is seen on the surface and in the shaft; a sample of 20 tons of the best ore from this shoot, taken from the tunnel, gave: gold, 0·03 ounces; silver, 10 ounces; copper, 8 per cent. Solid arsenical iron assayed: gold, 0·14 ounce; silver, 10·5 ounces.[1]

Table of Paragenesis of the Ores from the Cap Group.

Greenish tuffs and andesitic flows
Fissuring with some grinding
Sericitisation

[1]Assays: British Columbia Bureau of Mines, 1915, 1917.

Str... ...ification and deposition of some vein quartz
Magnetite
...yrite
Arsenopyrite
Sphalerite
Tetrahe...
...unite
Unknown mineral resembling enargite
Chalcopyrite
Galena

Chalcopyrite and arsenopyrite are the most abundant minerals, the others are present only in small amount.

The shoots are small and are separated by barren material, but the veins are strong and there is no doubt that they contain considerable ore.

GOLDEN WONDER PROPERTY. *General Description.* The Golden Wonder group is situated about 4,000 feet west of the Cap group and 1,200 feet lower; the elevation of the collar of the shaft is 1,290 feet.

There are three veins on the property, the north, centre, and south veins; only the centre has been worked to any extent.

The north vein is 12 to 15 inches wide and strikes north 85 degrees east (magnetic) with a dip of 75 degrees northwest. It follows the foot-wall of a 23-inch dyke of diorite porphyry which is also considerably mineralized with chalcopyrite, and has a 5-inch vein on the hanging-wall similar to the one on the foot-wall.

The vein is filled with ground-up country rock together with quartz and tourmaline, containing pyrite, chalcopyrite, arsenopyrite, and a little pyrrhotite.

The centre or main vein is 30 inches in width, strikes north 40 degrees east magnetic, and dips 80 degrees northwest. Besides stripping, a shaft is being sunk on the vein and was down 32 feet when examined. Near the top there was a considerable amount of chalcopyrite in a solid mass, but this passed into lower grade material much higher in arsenical iron carrying a little gold. Work is being continued to give the property a fair test.

The southern vein is exposed near the road and parallels the centre vein at about 300 feet distance. A shaft sunk 25 feet shows approximately 2 feet of pyrrhotite with a small amount of chalcopyrite which carries a little gold; the shaft has been abandoned, but the showing warrants further investigation.

Paragenesis. The tuffs here have been fissured and highly sericitized near the fissures. Silicification followed and quartz replaced nearly everything except the sericite. Tourmaline then replaced the quartz, so that both vein material and wall rock are high in tourmaline.

Table of Paragenesis of the Ores from the Golden Wonder Property.

Magnetite
Arsenopyrite
{Pyrrhotite
{Chalcopyrite
Marcasite
A later introduction of quartz was followed by the formation of
Siderite
Limonite
The exact position of the marcasite in the paragenesis could not be determined with certainty.

Properties on the North Side of Rocher Déboulé Mountain.

On the north side of Rocher Déboulé mountain the granodiorite forms the whole face of the mountain and there are several mineral locations on which very little work has been done, and which show very little promise. The Daley West property has been opened to the greatest extent, so a brief description will be given.

THE DALEY WEST PROPERTY. *General Description.* The Daley West property is situated about 2 miles southeast of New Hazelton and is reached by a good wagon road. It is about 3,500 feet from the northern contact and 6,500 feet from the eastern contact of the granodiorite with the sediments, and consists of a mineralized shear zone in the granodiorite.

The shear zone strikes north 15 degrees east and dips 65 degrees northwest; it is from a few inches to 3 feet in width and is traversed by a vein of quartz which carries the sulphides. The vein is exposed for about 500 feet up the hillside from the lower tunnel which is at about 1,675 feet altitude, and a second tunnel is driven 125 feet above the lower one.

The lower tunnel is 155 feet long and follows two small stringers which are 2 to 3 feet apart and separated by partly decomposed granodiorite; farther in the tunnel the stringers join and then separate as before. Each of the stringers carries an inch or two of chalcopyrite.

The upper tunnel is driven on the vein for 300 feet and carries a little chalcopyrite for about 100 feet in a vein up to a foot in width; the vein is then practically barren until near the face when a little chalcopyrite is again encountered, this time accompanied by some pyrrhotite which is not in evidence either on the surface or in the rest of the tunnel. The principal mineralization in the vein is of arsenopyrite carrying a very small amount of gold. The surface showings are not any more promising than those in the tunnel, and in view of the very low assays in gold the property does not appear to warrant any further expenditure.

In Bulletin No. 2, 1907, of the British Columbia Bureau of Mines, the following assays are given.

An average sample of about a ton of ore taken from the tunnel, mainly arsenical iron, gave: gold, 0·1 ounce; silver, 1·5 ounces; copper, 0·9 per cent.

A sample of solid arsenical iron gave: gold, 0·1 ounce; silver, 1·7 ounces; cobalt, nil; nickel, nil.

Properties on the East Side of Rocher Déboulé Mountain.

BLACK PRINCE GROUP. *General Statement.* This group is situated at the head of Mud creek near the eastern border of the main granodiorite mass, and is reached by a good trail from the New Hazelton-Telkwa wagon road. The trail is about 6 miles in length and leaves the road just north of the Grand Trunk Pacific Railway crossing. There are two main occurrences on the property, which are exposed on the face of a steep hill and are opened by cuts and short tunnels.

Description of Deposits. The first occurrence is apparently a shear zone in the granodiorite, with a maximum width of about 8 feet, definite walls, and containing one to four parallel stringers of quartz. The material in the zone is crushed and altered granodiorite which is considerably mineralized. The zone strikes north 62 degrees west (magnetic), dips

74 degrees southwest, and is opened up through an elevation of 550 feet and over 600 feet horizontally.

There are three principal cuts on the deposit; the lowest, at 4,500 feet, shows pyrite and molybdenite as being rather abundant, with a little chalcopyrite; the middle, at about 4,750 feet altitude, shows more molybdenite and less pyrite and chalcopyrite than the lower, together with a little wolframite. The material here is very highly leached.

The upper showing on top of the ridge, at 5,050 feet elevation, is of highly leached, rusty quartz with a considerable amount of wolframite, in lumps up to 4 inches in diameter.

The following assays were taken from the Annual Report[1] of the Minister of Mines for British Columbia, 1916.

"A sample of the vein-filling from an elevation of 4,600 feet gave: gold, trace; silver, 0·6 ounce; tungstic oxide, 4·0 per cent. The vein at this place is about 8 feet in width, but is mineralized only in stringers; four stringers are from 3 to 11 inches in width."

According to J. D. Galloway, the sample was taken across 4 feet of the vein, and the sampler was careful to exclude any of the lumps of wolframite, so that there was no visible wolframite in the sample taken.

"A sample across 2½ feet of leached vein material at elevation 4,650 feet, gave:

"Gold, trace; silver, 0·6 ounce; tungstic oxide, 1·1 per cent. At this place the stringers have come together again to form a single vein 2½ feet in width, highly leached and decomposed.

"A sample across 2½ feet of badly leached vein matter at an elevation of 4,670 feet, gave:

"Gold, trace; silver, 6·4 ounces; tungstic oxide, trace."

The second main occurrence is a chalcopyrite-hornblende vein somewhat similar to the Rocher De Boule veins. The vein is 8 inches to 14 inches in width and strikes north 65 degrees west with a dip of 53 degrees southwest. It is opened by a 50-foot adit tunnel at an elevation of 4,295 feet and extends up a steep cliff, but was not examined farther than the tunnel.

In the tunnel the vein is richly mineralized with chalcopyrite, pyrite, pyrrhotite, and molybdenite in a gangue principally of actinolite, with some quartz, siderite, and calcite. In places there are solid masses of chalcopyrite and pyrrhotite 12 inches in width.

Table of Paragenesis of the Ores from the Black Prince Group.

Wolframite
Unknown transparent mineral
Quarts and actinolite
Magnetite }small in amount and local
Hematite
Molybdenite (position approximate.)
Pyrite and pyrrhotite (?)
Chalcopyrite
Limonite

Plate VIIIB shows the replacement of wolframite by the unknown mineral. These are deep vein minerals and the deposit should extend to considerable depth; it is well worth investigating.

OTHER PROPERTIES. South from the Black Prince claims other, claims have been staked near the contact, usually on the sedimentary side; but none of them have yet been shown to be of importance.

[1]The elevations quoted were obtained by aneroid during an all-day trip, and do not necessarily agree with those of the present report which were from a plane-table traverse.

The sediments are usually highly pyritized for several hundred feet from the contact and in the neighbourhood of small dykes which are of frequent occurrence, and it is in this area that the claims are mostly located. The contact dips to the east at 50 to 60 degrees along this side of the mountain and better prospecting ground would probably be in places where it becomes flatter, that is, where the sediments are still in place on the sloping roof of the batholith rather than deep on the eroded sides.

MINING PROPERTIES NORTH OF BULKLEY RIVER: SILVER-LEAD-ZINC
PROPERTIES.

Properties on Glen Mountain.

SILVER STANDARD MINE. The Silver Standard mine is situated on the northwest side of Glen mountain, about 6 miles by road north of New Hazelton station, from where the ore is shipped. This mine started to ship in 1913 and up to June, 1917, had shipped 2,229 tons of silver-lead ore carrying 746,259 pounds of lead, 516·8 ounces of gold, and 304,411 ounces of silver, with an average of 20·3 per cent of zinc. In 1916 and to the end of May, 1917, 393·9 tons of zinc ore were shipped, which averaged 43·16 per cent of zinc, 0·24 ounce of gold, and 60·02 ounces of silver, making a total of 328,050·5 ounces of silver. During the latter part of 1917 the mine was shut down and all energies devoted to putting in a 50-ton mill to handle the large amount of second grade ore opened up in the development of the high grade shoots; this mill will be running in the spring of 1918.

General Description. There are nine veins on the property (*See* Map 1733), but only two of them have been important producers so far, although some of the others are known to contain high grade ore as well as the low grade material which is common to all the veins. The veins are number consecutively from west to east across the mountain. The strikes va tail, partly due to faulting, but the veins are roughly parallel an north 20 degrees to 35 degrees east (magnetic) with steep dips to the southeast.

The bulk of the ore taken out to date has been obtained from the main vein which occurs between veins Nos. 6 and 7.

About the middle of the property this main vein splits into two parts with apex to the south and an angle of about 10 degrees between the strikes of the two parts; these parts are known as the Hang-wall and Foot-wall veins respectively (Figure 5).

A shaft has been sunk for 400 feet on the Foot-wall vein about 250 feet north of the intersection, and levels have been developed at 150 feet, 250 feet, and at 400 ' from the top of the shaft. A crosscut tunnel was driven to intersect the main vein at the 250-foot level, 300 feet south of the shaft, and then from near the shaft a crosscut was driven 360 feet to intersect No. 7 vein. The tunnels intersect all the veins except Nos. 1, 2, and 8; these veins are only known from surface cuts and strippings.

Description of Veins. No. 1 vein has been proved for 1,500 feet by open-cuts along the strike and an ore-shoot 100 feet long by 20 inches average width gave an average sample assaying: gold, 0·35 ounce; silver, 10·72 ounces. The samples and assays are shown on the table page 28.

This vein contains galena and sphalerite in rich bunches, but is generally mineralised with arsenopyrite and pyrite with considerable siderite in a quartzose gangue, and carries much higher gold values than any of the other veins.

Analyses from No. 1 Vein, Silver Standard Mine.

No.	Horis. dist. from previous sample	Width of vein	Length of sample	Gold in ozs. a ton	Silver in ozs. a ton	Remarks
1	North end.........	4 ft.	....	0·42	12·36	Side of shaft, 12 ft. below surface.
2	9 ft. south........	28 in.	....	0·20	14·45	Side of shaft, 12 ft. below surface.
3	12 "	30 "	2 ft.	0·14	2·12	Outcrop.
4	5 "	21 "	2 "	0·23	29·31	Mainly solid zinc blende
5	4 "	2 ft.	2 "	0·54	6·34	Some solid zinc blende
6	4 "	12 in.	2 "	0·16	5·28	Some solid zinc blende (outcrop)
7	2 "	2 ft.	2 "	0·90	14·72	13 ft. below surface
8	0	2 "	4 "	0·10	2·74	19 ft. below surface
9	Contiguous south....	14 in.	4 "	1·00	13·00	Outcrop; mainly solid zinc blende
10	4 ft. south........	11 "	4 "	0·80	23·40	Outcrop; mainly solid zinc blende
11	5 "	12 "	4 "	0·50	13·10	Outcrop
12	5 "	12 "	4 "	0·75	26·95	Outcrop
13	5 "	6 "	5 "	0·30	9·80	Outcrop
14	250 "	12 "	....	0·04	7·50	Separate cut

Picked zinc ore from dump: gold, 0·15 ounce; silver, 20·45 ounces a ton.
Picked mixed ore from dump, with much arsenopyrite: gold, 1·10 ounce; silver, 19·90 ounces a ton.
Average of all samples, length 100 feet, width 1·7 feet: gold, 0·35 ounce; silver, 10·72 ounces a ton
NOTE. These analyses were furnished by the superintendent of the mine.

No. 2 vein is 100 feet distant from No. 1 and outcrops about 50 feet vertically above it. It has been opened for 300 feet and shows a vein 6 inches to 1 foot in width, carrying some ore.

No. 3 vein was the first encountered in the main crosscut tunnel, at 140 feet from the portal. It here showed about 6 inches of mixed ore.

No. 4 vein has been exposed at intervals along the strike for a distance of over 1,000 feet, and by a shaft 35 feet in depth. Ten tons of ore sorted from this shaft ran 289 ounces in silver. This vein has a sharply marked hangwall and the ore ranges from 12 inches to 3 feet in width.

The vein was intersected by the main crosscut tunnel at 410 feet from the portal, and drifting has opened an ore-shoot 110 feet long pitching 40 degrees to south, with the face still in ore. Samples along the drift, intended to be an average, were taken by the company, and the values are given on page 29.

In stoping on this vein conditions were found to remain constant, the northern half of the shoot being almost solid sphalerite, whereas the southern end was mixed ore or was high in galena and tetrahedrite.

The gangue is milky quartz and it has been crushed and sheared so that the ore is in veinlets as well as in massive replacements. Cross fissures filled with milky quartz are cut off by the main vein carrying the metallics.

Along one or both of the walls there is frequently a band of siderite with arsenopyrite, the central portion of the vein containing the silver-lead minerals. At one place an 8-inch vein of the former minerals parallels and in places touches the silver-lead vein, but was not seen to cut it.

Assays of Ore from No. 4 Vein.

Assays—250-foot level.	Gold	Silver	Lead	Zinc.
	Oz.	Oz.	Per cent	Per cent
118, W.—6-in. to 1 ft. from face to 20 ft. north of tunnel..	0·10	106·2	1·0	16·70
119, W.—6-in. to 2 ft. tunnel to 14 ft. south of tunnel....	0·62	338·00	9·57	30·00
120, W.—2-ft. from 119 to face........................	0·08	43·2	....	25·52
No. 4 vein assays (Averages).				
Car sample south of drift on 250 ft. level, 10-ft. advance.	0·04	31·6	1·86	7·52
95 sacks of zinc fines..............................	0·16	115·0	....	37·06
67 sacks of zinc fines..............................	0·20	99·0	....	33·79
320 sacks from stope 250-ft. level..................	0·18	99·0	3·34	35·95
463 sacks from stope...............................	0·20	100·5	....	37·71

No. 5 vein was not known on the surface until it was cut by the main tunnel, and projected upwards. The tunnel encountered the vein at 490 feet from the portal and exposed 12 inches of solid ore, mostly sphalerite and grey copper. This ore did not extend north of the tunnel, but was drifted on for 45 feet to the south. A general sample from this shoot, 12 inches across and 35 feet in length, gave the following assay:[1] gold, 0·16 ounce; silver, 106·60 ounces; lead, 1·59 per cent; zinc, 45·5 per cent; this is exceptionally high grade zinc ore.

On the surface the vein is small, but has been traced for a considerable distance. There are 155 feet of backs from the tunnel level to the surface, but it is not known how much ore is to be found there.

No. 6 vein was encountered in the main tunnel at 635 feet from the portal where a vein 12 inches to 2 feet wide contains a little sphalerite, galena, pyrite, arsenopyrite, tetrahedrite, and pyrrhotite, scattered through a matrix of milky quartz. There is no shipping ore exposed, but no drifting has been done as yet. A sample across the whole vein at the point of intersection of the tunnel gave:[1] gold, 0·02 ounce; silver, 2·4 ounces; zinc, 6·5 per cent; a low grade milling ore.

On the surface the vein had been known at an exposure 225 feet south of the centre line of the tunnel, where it is 6 to 14 inches in width and gave the following assay:[1] gold, 0·10 ounce; silver, 179·04 ounces; lead, 23·2 per cent.

The Main vein strikes north 25 degrees east, dips 70 degrees southeast, and is crosscut by the main tunnel at a distance of 870 feet from the portal and 290 feet south of the main shaft. Altogether there has been 500 feet of drifting on the Hang-wall section and 300 feet on the Foot-wall section of the vein. Besides the main shaft there are two raises connecting this level with the 150-foot level.

South of the tunnel, drifting was carried for 85 feet in a quartz vein with very little ore; to the north there were several small bunches of ore,

[1] Assays furnished by the mine.

but mostly barren quarts for nearly 200 feet or until the shoots developed by the main shaft were reached. There are shoots in both the Foot-wall and Hanging-wall veins near the shaft and they have been the chief sources of ore shipped to date.

The Foot-wall ore-shoot is known from the surface down to the 400-foot level and in a winze 86 feet deep from that level; it occurs at and near the intersection of the Foot-wall and Hang-wall veins. Shipping ore occurred in the shoot from the 150-foot level to the surface, where it varied from 6 inches to 8 feet in width, averaging 4 feet to the surface. The shoot at the 150-foot level is 200 feet long, at the 250-foot level about 150 feet long, with the same average width bunches of high grade in milling ore; at the 400-foot level it is about 100 feet long, varying from 18 inches to 3 feet in width, and all of milling ore. In the drift at the bottom of the winze, 86 feet below the 400-foot level, there are 6 inches of good grade ore for the total distance of 400 feet drifted. The manager reports as follows: "Samples taken down the winze at intervals assayed in silver from 12 ounces per ton up to 90 ounces per ton for widths from 2 to 6 feet, with minor percentages of lead and zinc, indicating the fact, which was quite apparent as we went down, that the ore contained more grey copper and less zinc and lead than in the levels above. An average sample of the fines at 40 feet in depth assayed as follows: gold, 0·12 ounce per ton; silver, 76·12 ounces per ton; lead, 6·6 per cent; zinc, 10·3 per cent."

The Hang-wall ore-shoot extends from the intersection of the Hang-wall and Foot-wall veins towards the north on the Hang-wall vein. At the surface this shoot was solid ore, mostly galena and grey copper, 12 inches in width and 40 feet in length; shipments from cuts at the surface averaged: silver, 175 ounces per ton; lead, 21·66 per cent; zinc, 14 per cent. On the 150-foot level the shoot is known to be 95 feet in length, but it has not been drifted on yet. On the 250-foot level the shoot is very prominent and shows 80 feet of continuous ore varying from 2½ to 4 feet in width; there is a considerable amount of shipping ore mixed with the main body of milling material. On the 400-foot level this shoot has been opened up in a drift over 100 feet in length and the ore is 6 inches in width, occurring along a well-defined hanging-wall. A sample taken over the whole length and a width of 6 inches gave the following values: gold, 0·234 ounce per ton; silver, 34·64 ounces per ton; lead, 2 per cent; zinc, 2 per cent (mine samples and assays).

A raise was put up from the 400-foot level in this shoot and it widened in a short distance to 3 feet, with a larger content of galena.

It will be noted that the relative proportions of gold and silver are higher than in upper levels.

No. 7 vein is situated about 400 feet from the main vein and is exposed at the surface, where a remarkable showing of ore 80 feet in length and 18 inches wide was discovered. Analyses[1] of the ore from this occurrence are as follows:

[1] Analyses furnished by mine.

Analyses of Ore from No. 7 Vein.

—	Gold Ozs. per ton	Silver Ozs. per ton	Lead Per cent	Zinc Per cent
18 feet below surface, 18 inches wide..............	0·26	183·0	17·1	30·0
5 feet from above sample, 20 inches wide........	0·08	107·20	13·0	40·0
36 sacks of ore.................................	0·30	224·3	24·8	27·7
50 sacks of ore.................................	0·24	229·8	16·3	30·3
28 sacks 18 inches to 2-ft. of sincy ore............	0·18	135·00	9·5	42·6
Shaft 16 feet deep contains 3·5 per cent of Cu.....	0·20	241·5	17·0	58·0
Average 35 sacks 18 inches to 2-ft.-vein..........	0·20	236·8	32·7	30·8
Average 106 sacks.............................	0·16	180·1	21·0	28·2

The zinc content of this shoot is very exceptional and the high silver and noticeable copper content suggests that the grey copper, or properly freibergite, is also abundant.

On the 250-foot level a crosscut has been run from the main shaft to tap this vein. The vein was found to be of milky quartz with bands 2 inches in width on both hang-wall and foot-wall, and it contained some pyrite, arsenopyrite, chalcopyrite, and siderite, but no galena, sphalerite, or grey copper; analyses gave only a trace of lead. The vein at this level varies in width from 6 inches to 3 feet, and in 170 feet of drifting no ore was encountered.

No. 8 vein outcrops about 100 feet vertically above and 400 feet horizontally east of No. 7 vein. Only surface work has been done on this vein so far, and it has opened an ore-shoot 100 feet in length and 3 feet in width. A crosscut from the 250-foot level would give 500 feet of backs on this vein.

There are several other veins east of No. 8 and at higher altitudes, but so far as is known at present they are of minor importance.

On the east side of the mountain an adit tunnel is driven in on a 15-inch quartz vein which strikes east and dips 60 degrees to the north. The vein has many small stringers branching from it into the sediments, and at 30 feet from the portal a fault zone filled with gouge and mineralized, cuts off the vein, apparently offsetting it towards the south. The fault strikes north 10 degrees east and dips about 80 degrees to the east.

Paragenesis of the Ores. The mineralization of the vein was similar to that in the veins described above, but pyrite is most abundant and the other minerals occur only sparingly in the part opened up.

Table of Paragenesis of the Ores from the Silver Standard Mine.

Fissuring of the tuffaceous sediments
Deposition of vein quartz
Siderite replacing vein quartz

Metallics—First mineralisation
{ Pyrrhotite
Pyrite } Positions varying, but all replace siderite and quartz
Arsenopyrite
Freibergite? (Small amount)
Chalcopyrite
Gold — Position later than arsenopyrite

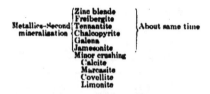

Metallics-Second mineralisation

Zinc blende
Freibergite
Tennantite
Chalcopyrite
Galena
Jamesonite
Minor crushing
Calcite
Marcasite
Covellite
Limonite

About same time

Discussion. The Silver Standard ore is very similar to the silver-lead ore of the Rocher De Boule mine, but it also contains jamesonite, a considerable amount of pyrrhotite, and a higher proportion of arsenopyrite. The order of deposition of the two latter minerals is interchangeable, and the veins of each cut the other in the same polished section.

The amount of zinc blende and tetrahedrite is much greater, and of galena somewhat greater than in the Rocher De Boule occurrence, thus giving a higher grade silver-lead-zinc ore. Plate IXA shows the average high grade ore. Free gold was observed in the arsenopyrite of the Silver Standard veins and jamesonite is commonly associated with the galena, in places in considerable amount.

Of the gangue minerals, calcite is abundant and is only slightly exceeded in amount by the quartz; the calcite is later than the quartz but preceded the metallic minerals as a rule, although small veins of calcite occur as the last phase of the mineralization.

There is very little chalcopyrite present and the values in copper are negligible.

The deposits may be classified as true fissure replacement veins. Galena, zinc blende, and freibergite are the most important minerals, but the arsenopyrite is important since it apparently acted as the precipitant for the gold.

Although two periods of mineralization are noted, there does not seem to have been any decided time-break between them and the deposition was more or less progressive with a cooling of the mineralizing solutions. It would thus be expected that in depth the minerals of the first group would tend to become relatively more abundant th.n the lower temperature group. Variations in dip and in width of vein are noted and these seem to be due in part at least to the different layers of sediment through which the veins pass.

The only factor so far shown to be associated with the location of the ore-shoots is the occurrence of small fissures which join the main vein at a small angle and which are themselves mineralized for a short distance from the intersection; these fissures are observed to be quite numerous near the more valuable shoots and were probably more or less concerned in localizing the main ore-bodies.

Properties on Ninemile Mountain.

West Side.

AMERICAN BOY PROPERTY. *General Statement.* This property is situated on the southwest slope of Ninemile mountain, about 6 miles from New Hazelton, and is reached from that place by a good wagon road which passes about 1,100 feet below the property; a switch-back trail leads from the road up to the mine workings.

There are a number of parallel veins on this property similar to those on the Silver Standard which is approximately 2 miles due west. These veins strike roughly north 30 to 40 degrees east, and dip 40 to 70 degrees southeast; they are offset by a series of small faults, but not to an extent to cause inconvenience in development.

The country rocks are of the Hazelton series, similar to those at the Silver Standard, and there are a few small dykes of granodiorite-porphyry which were earlier than the mineralization.

Since this property was shut down and there were no mine plans available, it was not studied in the same detail as was the Silver Standard, which furnished good facilities for an examination and is very similar to the American Boy occurrence.

There are three main veins opened up on the property, and a fourth which may be an extension of one of the others. A good description of these veins is to be found in the report of the Minister of Mines for British Columbia, 1914, and the assays given below are taken from that report.

The veins are strong, true, fissure veins which vary from 6 inches to 3 feet in width.

The mine was shut down because the high grade material was not sufficiently concentrated in the veins to permit of economical mining unless the second grade material which had to be taken out during development was utilized. Since funds were not available to erect a mill the property was closed. Now that the Silver Standard mine has erected a mill and will accept custom ore, this property is being reopened and should become an important producer.

No. 1 vein contains an ore-shoot 90 feet long on the surface, consisting of galena and zinc blende, with lesser amounts of jamesonite, tetrahedrite, pyrite, and chalcopyrite, in a gangue of quartz. The sulphides frequently exhibit a banded structure. The vein has been developed by a 100-foot shaft with short drifts at the 27 and 50-foot levels, and by a 25-foot shaft a short distance from the main one. A sample across 20 inches on the 27-foot level assayed: gold, 0·10 ounce; silver, 47·4 ounces; lead, 11 per cent. Picked high grade ore from the 25-foot shaft assayed: gold, 0·10 ounce; silver, 681·2 ounces; lead, 31 per cent.

No. 2 vein is 150 feet southeast of No. 1, somewhat similar in mineralization, and is 2 to 3 feet in width. A sample of selected ore assayed: gold, 0·15 ounce; silver, 481 ounces; lead, 41·2 per cent. There is said to be a shoot 120 feet in length on this vein.

No. 3 vein is about 600 feet from No. 2; it varies from 1 to 3 feet in width, and has been opened by a 180-foot shaft with short drifts from the 100 and 150-foot levels. About 100 tons of ore shipped in 1912, taken mostly from this shaft, is said to have netted about $7,000.

Paragenesis of the Ores. The paragenesis of the ores on the Silver Standard property also holds good for this property.

North Side.

General Description. The deposits in this locality are associated with a small boss of granodiorite which has been exposed on the north face of the mountain (Map 1731). The granodiorite is exposed in the shape of a wedge with apex to the east; it has a total length of approximately 9,000 feet and a greatest width of 3,000 feet. By the partial development of cirques the upper part of the north face of the mountain has been given a serrated plan, and on the steep walls of the basins formed the different veins are exposed and some can be traced across shoulders from one basin to the next; the lower slopes are covered with talus, this is well shown in P'ate IXB.

In the most westerly basin there are several claims on which some development has been done, but they are lying idle at present. The veins here are in the sedimentary tuffs and are found at least 1,500 feet west of the contact with the granodiorite. They strike north 5 degrees east and dips range from 25 degrees to 65 degrees southeast. They are true fissure veins 1 to 2 feet wide, usually well mineralized and containing bunches and bands of high grade ore, but mostly of ore classed as second grade; the sulphides are zinc blende, jamesonite, galena, and tetrahedrite mostly and carry good values in silver. With good milling facilities these properties should be worked at a profit.

SILVER CUP PROPERTY. This property has had the most work done on it and will serve as an illustration.

There are four main drift tunnels giving a vertical range of over 1,000 feet, although it is not certain that the lowest occurrence is on the same vein as the upper ones. The upper tunnel at about 5,000 feet altitude (See Map No. 171) is 200 feet long on a vein 6 inches to 2 feet in width, averaging 10 inches, and in ore all the way.

A sample across a 6-inch paystreak gave[1]: gold, 0·02 ounce; silver, 45·4 ounces; lead, 33 per cent; zinc, 22 per cent.

The second tunnel, about 150 feet lower, is 100 feet long and shows 20 to 24 inches of mineralized vein throughout its length. At 40 feet a stope was made for a height of 25 feet, all in ore. A sample assayed[1]: gold, trace; silver, 116 ounces; lead, 41·6 per cent; zinc, 12·4 per cent. Selected solid sulphide assayed[1]: silver, 150–250 ounces; lead, 50–70 per cent.

The lowest tunnel, the Duchess, was driven 140 feet, but after 100 feet the vein fingered out; the strike here is at about 45 degrees to that in the upper tunnels and it is probably one of another system of veins.

The vein is 6 to 18 inches in width, with disseminated sulphides, mostly arsenopyrite, for 6 inches in the sedimentary tuffs on either side. The gangue is quartz and a sample across 18 inches gave:[1] gold, 0·06 ounce; silver, 92·2 ounces; lead, 14·9 per cent; zinc, 11·16 per cent.

SUNRISE PROPERTY. In the next basin, one-half mile to the northeast of Silver Cup, the principal showings are on the Sunrise property where there are two large mineral zones in granodiorite. The lower is at 4,950 feet elevation; it strikes north 10 degrees east and dips 45 degrees southeast. The zone is 20 feet in width and contains a clean vein 1 foot in width at the bottom and a narrower one at the top; the whole zone is well mineralized, chiefly with galena and jamesonite. It has been opened for 200 feet. The upper zone is at 5,050 feet elevation; it is 6 to 8 feet in width and is highly mineralized throughout, chiefly with galena and jamesonite. It has been opened by cuts for 100 feet and strikes north 60 degrees east with a dip of 40 degrees southeast. Near the middle of the zone there are 3

[1] Assays by British Columbia Bureau of Mines.

to 12 inches of clean sulphide, consisting mostly of finely banded galena and jamesonite; near the top there is a band of 8 inches, mostly of zinc blende. These zones have good walls and are said to carry high silver values; the showings are well worth further development and are exceptionally promising.

Assays by the British Columbia Bureau of Mines of samples from Sunset group, are as follows:

Bright clean galena from a 4-inch stringer in mineralized shear zone: gold, 0·02 ounce; silver, 140 ounces; lead, 72·5 per cent.

Across 5 feet of shear zone: silver, 49·8 ounces; lead, 24·2 per cent; zinc, 5·2 per cent.

Picked ore from same zone as No. 2: gold, trace; silver, 86 ounces; lead, 61·6 per cent; zinc, 7·5 per cent.

Farther east in the other basins similar deposits are known, but the above descriptions are typical.

Paragenesis of Ores. The paragenesis of the ores from this locality is similar for all the deposits and is as follows.

Table of Paragenesis of the Ores on the North Side of Ninemile Mountain.

Fissuring of the country rock with the formation of veins and shear zones
Deposition of quartz with some calcite
Siderite
Arsenopyrite and pyrite
Zinc blende
Tetrahedrite ⎫
Galena ⎬ veining zinc blende
Jamesonite ⎭

Zinc blende is usually the most abundant metallic mineral, then jamesonite, and then galena; there is considerable finely crystalline arsenopyrite, but not so much of the other minerals. Plate XA shows the relationship of the jamesonite to the zinc blende.

Discussion. The Ninemile ore consists only of the following minerals: pyrite, arsenopyrite, zinc blende, galena, jamesonite, and tetrahedrite, in a gangue of quartz and calcite. The pyrite and arsenopyrite are distinctly earlier than all the other minerals and are followed by zinc blende and tetrahedrite (jamesonite and galena). The calcite is earlier than the metallic minerals and is commonly found thickly impregnated with needles of jamesonite as in Plate XB. There does not appear to have been two distinct periods of mineralization, but simply a gradual cooling of the solutions as they precipitated their minerals in these veins.

It seems unwarranted that these properties should remain idle at a time when lead and silver are in such demand. They could be reasonably proved at no great expense and they are admirably situated for co-operative handling of ore to a mill-site, say on the Shegunia river just below them; this stream could also furnish all the power necessary for a large development. At present there is a good wagon road from the railway at New Hazelton to the properties, a distance of about 10 miles by road.

Properties on Fourmile Mountain.

Fourmile mountain is situated about 4 miles northeast of Hazelton, just north of the Bulkley river, and is reached by a good wagon road. It rises to 2,200 feet altitude, or 1,400 feet above the river, and has an almost circular core of granodiorite which is about 4,000 feet in diameter.

On the north, west, and south sides of this mountain usually in the sediments near the contact, mineral claims have been located and considerable work done on them in the shape of trenches, drifts, and short shafts. None of these properties are shipping and only a slight amount of work is being attempted on any of them.

The ore occurs in quartz veins in shear zones which are several feet in width, with strikes varying from north 10 degrees east to north 45 degrees east, and steep dips to the southeast. The sulphides are galena, sphalerite, jamesonite, grey copper, and pyrite. These carry high values in silver,[1] as seen from the following analyses of ore from the Era (probably the Erie group) taken across 18 inches to 2 feet of solid sulphides.

(1) Gold, 0·08 ounce; silver, 358·17 ounces: lead, 7·81 per cent; copper, 0·75 per cent.

(2) Gold, 0·02 ounce; silver, 46·16 ounces; lead, 6·90 per cent; copper, 0·26 per cent.

The high grade ore is apparently bunched or occurs in small shoots accompanied by second grade ore and there was not enough of the former to warrant continued development when the latter could not be utilized.

Though none of the showings examined were of importance, some of them gave promise of value and in view of the present high price of silver, together with the possibility of using the Silver Standard concentrating mill, it seems that the best of these deposits warrant considerable attention. The possibilities in this district should be as good as those at the American Boy or at the Silver Standard, except that jamesonite is relatively more important in these deposits and it does not contain as much leadyas the galena of the other properties.

NON–METALLIC DEPOSITS.

CLAYS.

The clays in the district under consideration are chiefly "gumbo" or boulder clayyand have no economic importance; where the gumbo has been re-sorted by river action, some good clays have been deposited. A sample of clay was collected by J. D. MacKenzie of the Geological Survey, from the east bank of the Bulkley river, 300 yards south of the bridge at Smithers, and a similar sample from the same general locality was submitted for testing by Mr. Duke Harris; these samples were investigated by J. Keele of the Ceramic division of the Department of Mines, Ottawa, who reports as follows:

"Stratified, grey, silty, non-calcareous clay on the east bank of Bulkley river, 300 yards south of bridge at Smithers, B.C.

"This clay requires 40 per cent of water to bring it to the best working consistency. It is rather sticky when wet but works up into a smooth body which flows fairly well through a tile die.

"The shrinkage on drying is 10 per cent which is excessive. It burns to a good, hard, red body at 1,800 degrees F. If burned to a higher temperature than this the shrinkage becomes too great.

"The clay is easily fusible and cannot be used in the manufacture of vitrified wares.

[1]Leach, W. W., Geol. Surv,. Can., Sum. Rept., 1909.

"Its uses are confined to the manufacture of common building brick and field drain tile, but it would require the addition of 20 to 30 per cent of sand in order to reduce the shrinkages.

"The sample submitted by Mr. Harris is practically the same kind of clay as that at Smithers. It would be suitable for making common brick if sand could be found convenient to the deposit. The addition of not less than 25 per cent of sand would be required for brick-making."

SAND.

Near Hazelton, on the north bank of the Bulkley, just east of the bridge, there are terraces of well-assorted sands, but no detailed examination has been made of them.

TRAVERTINE.

Just above Twomile townsite, along Twomile creek, there are deposits of travertine where the groundwater has oozed out of the clay or formed springs; the lime carbonate is very pure in places and gives the appearance of extensive beds, but it is very probably of shallow depth, merely an outer coating on the clay and grading into the ordinary boulder clay of the district. The groundwater in passing through material containing calcium minerals has become highly charged with lime which was precipitated as the carbonate when the waters reached the surface.

PLATE II.

A. Bulkley canyon at Hagwilget; looking south towards Rocher Déboulé mountains. Note the change in topography at the upper limit of valley glaciation. (Pages 2, 4.)

B. Tuff-agglomerate in a cut in the Grand Trunk Pacific railway, near Beament station. (Page 4.)

40

Plate III.

A. Interbedded fine-grained tuffs and tuff-conglomerates in a cut on the Grand Trunk Pacific railway, north of Beament station. (Page 4.)

B. Sharp contact of the sediments with the granodiorite on the ridge between the head of Juniper creek and the head of Balsam creek, Rocher Déboulé mountains. (Page 5.)

PLATE IV.

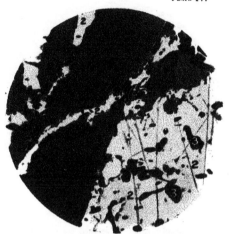

A. Microphotograph of polished ore from Rocher
De Boule mine, magnified 250 diameters. Chalcopyrite veining hornblende (first mineralization). 1,
hornblende; 2, chalcopyrite; 3, quartz. (Page 12.)

B. Microphotograph of polished ore from Rocher
De Boule mine, magnified 200 diameters. Hornblende
(actinolite) replacing chalcopyrite of the first mineralization. 1, magnetite; 2, pyrrhotite; 3, chalcopyrite;
4, actinolite; 5, pyrite. Paragenesis—1, $\overline{2, 5}$, 3, 4.
(Page 12.)

42

PLATE V.

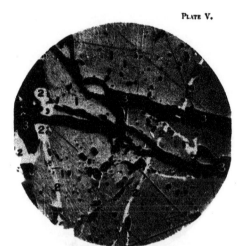

A. Microphotograph of polished ore from No. 2
vein of the Rocher De Boule mine, magnified 200
diameters. Chalcopyrite veining and replacing tetrahedrite, and both cut by veins of quartz. The sulphides are considered to be of the first mineralization
and the quartz is probably of the second. 1, tetrahedrite; 2, chalcopyrite; 3, quartz. (Page 12.)

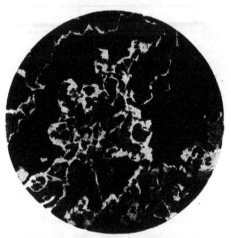

B. Microphotograph of polished ore from No. 2
vein of the Rocher De Boule mine, magnified 200
diameters. Arsenopyrite of the second mineralisation replacing chalcopyrite of the first. 1, chalcopyrite; 2, arsenopyrite; 3, quartz. (Pages 12, 13.)

PLATE VI.

43

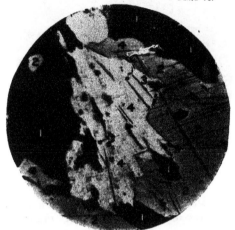

A. Microphotograph of polished ore from No. 4 vein of the Rocher De Boule mine, magnified 200 diameters. Molybdenite replacing chalcopyrite and hornblende. 1, hornblende; 2, chalcopyrite; 3, pyrrhotite; 4, molybdenite. (Page 13.)

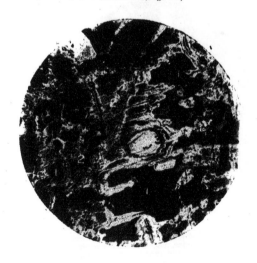

44

PLATE VII.

A. Microphotograph of polished ore from the Brian Boru property, magnified 395 diameters. Chalco-pyrite veining zinc blende. 1, zinc blende; 2, chalco-pyrite; 3, pyrrhotite; 4, gangue. Paragenesis—4, 1, 3, 2. (Page 20.)

B. Microphotograph of polished ore from the Hazelton View mine, magnified 395 diameters. Showing the fineness and distribution of the gold. 1, arsenopyrite; 2, safflorite; 3, gold; 4, molybdenite; 5, gangue (quartz, actinolite, etc.). (Page 23.)